	DATE DUE		

15485

660.6 Stanley, Debbie.
STA
Genetic engineering
: the cloning debate

MESA VERDE MIDDLE SCHOOL
POWAY UNIFIED SCHOOL DISTRICT

443347 01995 51236C 52097F   008

FOCUS ON SCIENCE AND SOCIETY

GENETIC ENGINEERING

The Cloning Debate

Cloning is the creation of an organism's genetic twin.

FOCUS ON SCIENCE AND SOCIETY

GENETIC ENGINEERING

The Cloning Debate

Debbie Stanley

The Rosen Publishing Group, Inc.
New York

GENETIC ENGINEERING: THE CLONING DEBATE

Published in 2000 by The Rosen Publishing Group, Inc.
29 East 21st Street, New York, NY 10010

Copyright © 2000 by The Rosen Publishing Group, Inc.

First Edition

All rights reserved. No part of this book may be reproduced in any form without permission in writing from the publisher, except by a reviewer.

Library of Congress Cataloging-in-Publication Data

Stanley, Debbie.
   Genetic engineering : the cloning debate / Debbie Stanley.
     p. cm. —(Focus on science and society)
    Includes bibliographical references and index.
    Summary: Examines the nature, history, and ethical aspects of cloning, discussing both humans and other animals.
       ISBN 0-8239-3211-7
    1. Cloning—Juvenile literature. 2. Cloning—Moral and ethical aspects—Juvenile literature. [1. Cloning 2.Genetic engineering.] I. Title. II. Series.

QH442.2 .S73 2000
660.6'5—dc21   00-021283

Manufactured in the United States of America

CONTENTS

	Introduction: What Is Cloning? 7
Chapter 1	Cloning of Animals 13
Chapter 2	Cloning of Humans 20
Chapter 3	The Ethics of Cloning 24
Chapter 4	Should Cloning Continue? 44
	Glossary 55
	For Further Reading 57
	For More Information 59
	Index 62

Scientists use an organism's DNA to create a clone.

INTRODUCTION: WHAT IS CLONING?

Cloning is a form of genetic engineering in which the DNA of a person, animal, plant, or even a bacterium is used to produce a perfect or near-perfect genetic replica of the original.

True cloning is the creation of a genetic twin of the original organism. This twin is younger than the original, unlike naturally occurring twins, but it carries the same genetic makeup as the original, as do naturally occurring twins. Technically, there is a slight difference in the genetic makeup of the clones that have been created to date, but the difference is so small that scientists still consider the cloned creatures to be genetic duplicates of the originals.

Scientists speculate that a clone would look exactly the same as the original and that it would experience the same consequences of its genetic makeup as the original—if the original got cancer because of a genetic defect, the clone would, too—but since no human clones have actually been

GENETIC ENGINEERING: THE CLONING DEBATE

created (or, at least, not reported), no one knows for sure. Still, anyone who has known a set of identical twins knows that even though you might have trouble telling them apart at first, after a while you would come to see their differences and even discover that they do not look exactly the same. Also, not every person whose twin has cancer also gets it. The differences between twins who share the same genetic makeup—who are, in other words, naturally occurring clones—make the task of predicting how much a clone would resemble its original even more difficult.

How Cloning Is Done

There are several different methods for creating a genetic duplicate of an original organism, most of which are variations on the same technique, but scientists racing to discover new

Timeline

1941–1945	1950s–1960s	1970s	1980s
World War II; the Allied forces defeat Adolf Hitler of Germany, whose plans to use eugenics to build an Aryan "master race" resulted in the murder of six million European Jews.	Science-fiction movies such as *Invasion of the Body Snatchers* speculate on the theme of human cloning.	The U.S. corn crop is lost due to monoculture (a restricted genetic base).	The science of cloning progresses in complexity only as far as the tadpole.

INTRODUCTION: WHAT IS CLONING?

methods or to perfect the old ones are very protective of the details of how they work.

In the past, the most widely used method of cloning was to take a cell from a developing organism, or embryo, before the cells had become differentiated. The process is, in essence, using an artificial method to create twins the same way they would occur naturally. The most important step is to divide the cells before they have become differentiated. Once a cell is differentiated, it is on its way to becoming a certain type of cell, such as blood or skin. Before then, however, each cell has the potential to create an entire organism. Cells that are not yet differentiated are called totipotent cells, while those that have become differentiated are called somatic.

The modern cloning method that has received the most attention in recent years is known as somatic cell nuclear transfer. The process involves removing the nucleus of a somatic cell and using that nucleus to replace the

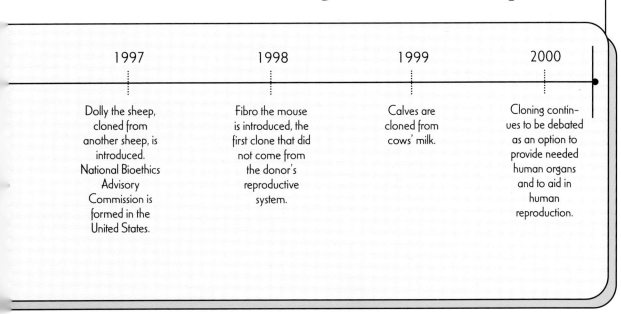

1997	1998	1999	2000
Dolly the sheep, cloned from another sheep, is introduced. National Bioethics Advisory Commission is formed in the United States.	Fibro the mouse is introduced, the first clone that did not come from the donor's reproductive system.	Calves are cloned from cows' milk.	Cloning continues to be debated as an option to provide needed human organs and to aid in human reproduction.

GENETIC ENGINEERING: THE CLONING DEBATE

Scientists are racing to discover new methods of cloning.

nucleus of an egg cell, or ovum. The resulting embryo will be, genetically, the same as the organism that provided the somatic cell nucleus. That embryo can then be implanted into the uterus of an animal that will later give birth to it, or it can be grown in a laboratory.

Cloning does not have to result in an entire animal. In many cases, cloning is used to create a "cell line," which is

INTRODUCTION: WHAT IS CLONING?

the result of cloning individual cells. This can be used to produce substances that are beneficial to humans, such as insulin. Much of the work being done in this area is not exactly cloning but is genetic engineering.

Genetic Engineering

Genetic engineering involves more than just the cloning of humans or animals. It is also used to create more perfect plants; the goal is to develop crops that are resistant to diseases or insects, or that produce more seeds, more or better fruit, or a bigger overall harvest. By tinkering with the genetic makeup of plants, scientists can encourage desired traits and eliminate unwanted ones. These traits may be things that natural selection would have removed because they do not support the continued existence

THE CLONING MYTH

The science-fiction image of a clone is of a person who looks, acts, and thinks exactly like the original person. This is a myth. A human clone would be genetically identical to the donor person but would be his or her own person, with a unique personality and capable of independent thought. The clone would not be a robot mimicking the donor's every word, thought, and action; it would not be the same person all over again. Still, just how similar the original and the clone would be is not yet known.

GENETIC ENGINEERING: THE CLONING DEBATE

of the species but that are beneficial for the farmers or the consumers.

Similar experimentation is being done with animals. Farmers have been practicing their own form of genetic engineering for thousands of years by using animals with desirable traits as breeders instead of as food. Modern genetic engineering eliminates the need for the farmer to wait until an animal with the right traits comes along.

Genetic engineering is used to create more perfect plants.

Another example of genetic engineering that is similar to cloning is the creation of animal hybrids (crosses between two species of animal) or of chimeras, which are combinations of animals and humans. The goal of these experiments is usually to find new ways of creating substances or organs that can be used in humans.

CLONING OF ANIMALS

CHAPTER 1

The announcement in 1997 that a newborn sheep named Dolly had been cloned from another sheep made worldwide news and threw scientists, governments, and people around the globe into a frenzy of debate over the moral implications of cloning. In actuality, experiments in cloning had been going on for at least twenty years, but none had grabbed the world's attention like this one. The difference with Dolly was that people began to believe in the possibility of cloning humans.

In the relatively short amount of time since Dolly's birth, many other cloning experiments have been conducted and some of them have produced more clones—mice and cows, for example. Other experiments have created animals with some human traits through the combination of animal and human DNA. Still others have combined different animals to create new ones. One such

GENETIC ENGINEERING: THE CLONING DEBATE

experiment crossed a goat and a sheep to produce what the scientists called a "geep."

Dolly's birth seemed to spark a worldwide race to create more clones and to do it in new and better ways. Since Dolly, many other female animals have been cloned, including a mouse named Cumulina and several Japanese cows. In 1998, a male mouse named Fibro became the first live mammal cloned from adult cells that did not come from the donor's reproductive system—he started out as part of another mouse's tail. Fibro, created at the University of Hawaii, has even fathered his own natural offspring, all of which appeared to be normal. In 1999, Japan reportedly

Dolly, the cloned sheep, made headlines in 1997.

succeeded in cloning calves from cells found in cows' milk. Other entrants in the cloning race have included cows and pigs from the United States, France, and Australia. Unconfirmed reports have also circulated that some South Korean scientists began the cloning of a human being but stopped the experiment in its early stages.

Dolly was created using somatic cell nuclear transfer, and she has so far lived what seems to be a normal life. Researchers have continued to study her, however, and have made further discoveries about her. For example, they found that, genetically, she is not 100 percent the same as the original she was cloned from.

Several years after Dolly's birth, the scientists who created her, Ian Wilmut and others at the Roslin Institute in Scotland, discovered that a subsection of her

... TRY, TRY AGAIN

From a strictly statistical perspective, it is amazing that Dolly was ever born in the first place. The experimentation that created her also included 276 failed attempts. In the case of Fibro the mouse, there were 273 failures. This extremely poor success rate and the fate of the creatures who were the "mistakes" have not been the most hotly contested aspects of the cloning debate. However, if these "failures" had involved humans instead of animals, public outcry would surely have been much louder. The potential for the destruction of human embryos in the course of cloning attempts was one of the biggest issues raised by legislators debating whether cloning should be made illegal in the United States.

GENETIC ENGINEERING: THE CLONING DEBATE

DNA, created not in the nucleus but by the mitochondria outside the nucleus, matched the egg cell, not the donor cell. This mtDNA constitutes a very small percentage of Dolly's overall genetic makeup, but it can cause unexpected genetic abnormalities. People who argue that the cloning of humans would be dangerous and would yield unpredictable results point to this discovery to support that opinion.

Further study of Dolly's genetic makeup has revealed that, genetically, she might be older than her chronological age. An article in the journal *Nature* indicates that the tips of Dolly's chromosomes, called telomeres, are shorter than they should be for her age. Shorter telomeres are believed to be an indication of a shorter life expectancy and could mean that Dolly will not live as long as other sheep her age—that, in effect, when she was born she was already several years old (the same age as the sheep she was cloned from). If this is true, it means that there is a limit to the number of generations that can be cloned. For example, if the first donor is five years old, the clone made from it will be five years old, genetically, at birth. If the clone is used to create another clone five years later, that second clone will be born ten years old, and so on. If it turns out that cloning does indeed result in premature aging, that will be another argument against the cloning of humans.

So far, no federal laws have been passed outlawing cloning research, but President Clinton did appoint a group

CLONING OF ANIMALS

of people to study the matter and make a recommendation. The National Bioethics Advisory Commission, formed in 1997 in reaction to the news of Dolly's creation, concluded that many other studies of cloning would be needed in order to determine if it is safe or effective. The commission recommended that no human cloning be attempted and that no federal funding be supplied for such experiments. "In its report," Ron Seely wrote in the *Wisconsin State Journal*, "the commission warned that the technique used to create Dolly was successful in only 1 of 277 attempts. Those odds haven't been substantially improved upon since then. As a result, the commission warned, there are real dangers

Fibro the clone mouse was introduced in 1998.

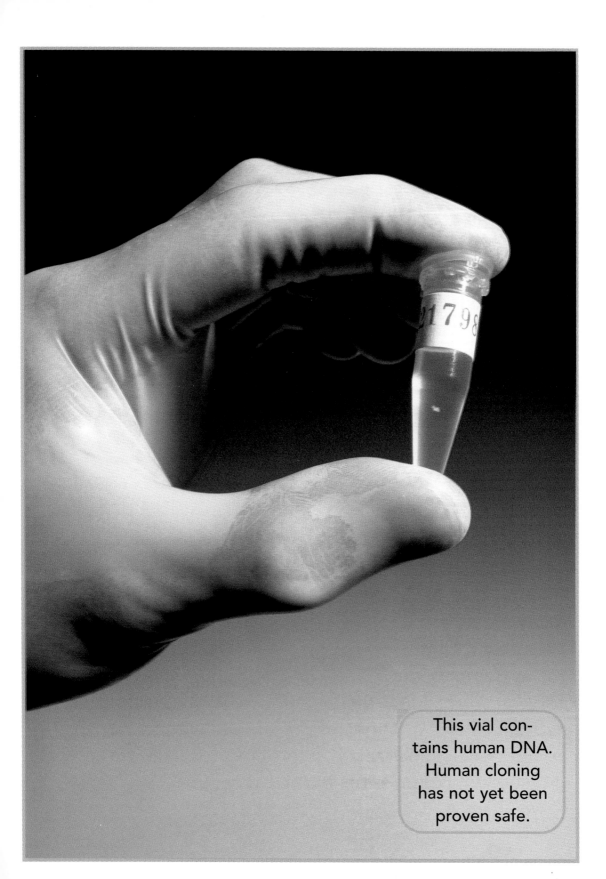

CLONING OF ANIMALS

if the techniques are attempted on humans. These include multiple miscarriages in the birth mother and possibly severe developmental abnormalities in any resulting child."

In light of the commission's conclusions, some groups have voluntarily agreed to stop experimentation in cloning. The American Society for Reproductive Medicine, for example, issued a five-year moratorium on cloning, stating that it has not yet been proven safe.

CLONING OF HUMANS

The idea of cloning humans is either scary or exciting for most people. So far, no one has created a cloned human, or at least no one has admitted to it, and there are no "cloning clinics" to go to if you want a clone of your own. However, with all of the experimentation that has been done on animals, many scientists believe that it would be possible to apply the same technology to humans. What the results would be, though, is anyone's guess.

Richard Seed, a Chicago scientist, brought arguments over cloning to a fever pitch when he announced early in 1998 that he and his brother intended to begin using cloning to create babies for infertile couples. Seed told National Public Radio: "It is my objective to set up a human clone clinic in greater Chicago, make it a profitable fertility clinic, and when it is profitable, to duplicate it in ten or

CLONING OF HUMANS

Scientist Richard Seed is a cloning advocate.

twenty other locations around the country, and maybe five or six internationally." Seed reportedly told NPR that he already had the needed equipment to begin trying the procedure and added, "We are going to have almost as much knowledge and almost as much power as God."

Seed enraged many members of the scientific community with what they saw as his flippant and arrogant attitude.

GENETIC ENGINEERING: THE CLONING DEBATE

Alta Charo, an ethicist and member of President Clinton's National Bioethics Advisory Commission, recalled that Seed had actually announced his plans a month earlier at a conference of researchers. Ron Seely of the *Wisconsin State Journal* reported that Charo was in attendance at that conference: "'At the time,' Charo recalled, 'he said he would do this because it would be so much fun.'" Seely noted that Charo said she and others "bristled" at Seed's mention of his plan, which he made as an audience member while asking a question of a speaker, and that on the day after Seed's announcement on NPR, Charo stated, "The National Bioethics Advisory Commission's concern is that children not be the unwitting victims of such fun…. With only a couple of sheep and a handful of cattle, we are not even close to having completed responsible testing on this technique. Whatever the Seed brothers are telling their human clients, they are certainly not able to tell them what the risks might be, since no one yet knows the range or magnitude of risks in primates, let alone humans."

What Could the Future Hold?

It would be impossible to predict what might happen with genetic engineering in the future. Remember, just a few decades ago, no one believed it was possible to clone an organism, and now we know that it is. However, there are

CLONING OF HUMANS

still many limitations. For instance, as of now it takes a living cell to make a clone. This means that it is not possible to clone someone who is already dead. There is also a very large failure rate—usually hundreds of failed attempts for each successful clone—but with practice the technique would likely become more reliable. Also, if cloning became more commonplace, the cost would likely go down.

The amount of time it will take to refine the cloning process will likely depend on whether the practice is condoned or condemned by society and the government. If people are for it, there will be more funding available to continue cloning; if not, funding will have to come from private sources and will be available only to those who can afford it. If it is outlawed, some people would probably continue to experiment in secret, risking fines or imprisonment.

One thing is certain: The belief commonly held just a few years ago that human cloning was either impossible or thousands of years away has been shattered. We now know that it can be done. The question is, will it be done?

THE ETHICS OF CLONING

There are many arguments for and against cloning, and many of them are discussed in this chapter. In addition to the two camps of cloning advocates and opponents, there is also a group of people who believe more research is needed before a decision can be reached on whether cloning should or should not be allowed. This third group occupies a middle ground between those who are for cloning (and therefore believe research should continue) and those who are against it (and therefore believe further research should be banned).

Arguments for Cloning

Some people who are in favor of cloning admit that there are risks and that the technology is not yet perfected, but they believe that the benefits outweigh the risks. Others

THE ETHICS OF CLONING

refuse to believe that there are risks at all. Arguments for cloning can be categorized as either individual or societal.

Benefits to Individuals

The cloning and other forms of genetic engineering being done on animals could lead to many health benefits for humans. Animals that have been bred with human genes could be used as subjects in drug testing, with the results being more applicable to humans than those of current tests. The breeding of these transgenic animals could also lead to cross-species organ transplants (called xenotransplantation), a breakthrough that would in theory solve the current problem of not having enough organs for

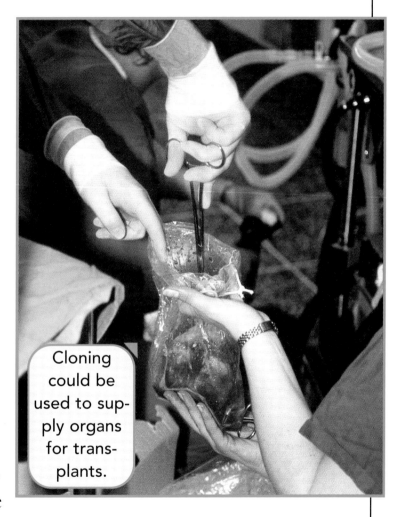

Cloning could be used to supply organs for transplants.

GENETIC ENGINEERING: THE CLONING DEBATE

everyone who needs them. This might even eliminate the black market for human organs that currently exists to provide organs to people who can afford to pay huge fees for them. It is interesting to note that the National Bioethics Advisory Commission did not recommend the banning of research on primates because they believed that such research could be beneficial to humans.

Human cloning could also be used to repair the health of other people. In a staff editorial reproduced by University Wire, the *Minnesota Daily* asserted: "Close to three thousand people die each year in America while waiting for donor organs. Human cloning could complement or provide an alternative to xenotransplantation.... Also, in organ transplants, it is common for the immune system to reject a foreign organ. Readily available cloned human organs could eliminate these problems."

Cloning would not have to be carried to the point of creating an entire animal. The cloning of individual cells, creating what is known as a cell line, is a technique already in use to create substances that aid people's health, such as treatments for diabetes, hemophilia, or cystic fibrosis. In some cases, these cell lines can be used by themselves, and in other cases, they must be introduced into host animals that then produce the needed substance—in their milk, for example. Another example is the biotech firm Prolume Ltd., which has cloned an enzyme in jellyfish and other sea animals that give off light, and is planning to use it to create a substance that would light up tumor cells and make them more visible to doctors.

One of the most compelling arguments for cloning is that it could become a new form of artificial reproduction for humans. Some see it as a long-awaited miracle solution for infertility, or as a way to ease the grief of losing a child, or even as a way of providing perfectly matched donor materials (organs, blood, bone marrow) for a sick child. People who are in favor of this use of cloning make up a very vocal segment of all cloning advocates.

Benefits to Society

All of the individual benefits of cloning could also be seen as benefits to society because they would allow people to live longer,

THE "NATURAL ORDER"

On a larger scale, many scientists are also very much against interfering with the "natural order." It is possible that cloning, depending on where, how, and how widely it is practiced, could create unpredictable and undesirable problems, such as imbalances in the food chain, which could lead to catastrophic damage to the world's ecosystem. Among humans, giving people the ability to live longer would mean a greater strain on the world's resources and on social services, utilities, and even housing. With more and more people being born, or created, and fewer people dying, the ability of our planet to sustain everyone could be compromised.

GENETIC ENGINEERING: THE CLONING DEBATE

healthier lives and therefore have a better chance at making a valuable contribution.

Besides helping individuals to live longer, more productive lives, some advocates of cloning also believe it could be used to create more people with traits that are especially valuable—people like scientists Albert Einstein, Marie Curie, and Stephen Hawking; artists Michelangelo and Picasso; authors William Shakespeare, Emily Dickinson, and Charles Dickens; politicians George Washington and Abraham Lincoln; and spiritual leaders Martin Luther King Jr., Mother Teresa, and Mahatma Gandhi. If the extraordinariness of such

Should people like Einstein be cloned?

THE ETHICS OF CLONING

people is recognized during their lifetimes, cloning advocates argue, they could contribute their genes to be cloned for the good of society.

While most of the benefits of cloning are for humans, there is at least one for animals. Cloning could be used to prevent species from becoming extinct. If a species of animal is nearing extinction because of a genetic defect, just one animal without that defect could produce many clones that could begin to rebuild the species. If the reason is simply that not enough animals are being born naturally, cloning could be used to bulk up the population. If cloning could be used to prevent extinction, some say that would benefit not just the animals but society as well.

Arguments against Cloning

Some people who are against cloning object to it simply because of the risks, not for moral reasons. These people might change their minds if cloning became more reliable. Others will never accept cloning, even if the technology is perfected, because they believe it is morally wrong. Arguments against cloning can be categorized as either scientific or ethical.

Scientific Objections

By far the strongest scientific objection to cloning is that the technology is not refined enough yet. With the relatively

A sheep's egg being injected with a cell during cloning.

limited amount of experimentation that has been done, scientists have learned how to make clones, but they have not learned what will happen to those clones over their lifetimes, or to any offspring the clones themselves might have. Not enough time has passed for that type of information to be revealed. Questions about the life expectancy of clones, their resistance to disease, the possibility of introducing new diseases into the human population, and the fate of babies that are born as imperfect attempts at cloning all have no answers at this point.

Some people fear that eliminating genetic differences among organisms through the use of cloning could be disastrous when disease strikes. The Council for Responsible Genetics, in its "Position Statement on Cloning," argues that a population of animals

CLONED CHILDREN

If cloning should become an accepted form of artificial reproduction, some worry about what will happen to the babies. Sharon Schmickle reported in the *Minneapolis Star Tribune:* "'Discussions of reproductive rights also should consider what's best for the children who are born,' said Gladys White, executive director of the National Advisory Board on Ethics in Reproduction. Scenarios of cloned children range from horror stories of organs being harvested from babies to calm predictions of each child, regardless of its genes, becoming a unique person shaped by time and events."

benefits from genetic variety because that variety allows individual animals to respond differently to diseases. If all animals in a population were affected the same way, they could all die of the same disease at the same time. "The robustness of natural populations, including their flexible response to new conditions and hence resistance to disease, lies to a great extent in their genetic variability. This characteristic would be entirely eliminated in a population of clones," the council states. "The near total loss of the entire U.S. corn crop in the 1970s as a result of monoculture—overuse of too narrow a genetic base—is a harbinger of what could happen with cloned livestock." Some believe this risk would also exist in human cloning.

Ethical Objections

Some opponents of cloning believe that turning animals into renewable resources lowers our respect for them and, by extension, for human life. The Council for Responsible Genetics believes that the creation of ideal "agricultural animals" through cloning will "inevitably undermine any respectful stance toward animals that may remain in our highly corporatized culture." In its "Position Statement on Cloning," the council asks, "Are we prepared to view animals solely as lucrative biofactories, useful only in their capacity to serve human needs?"

Some people object to the idea of using animals as bio-factories to serve human needs.

GENETIC ENGINEERING: THE CLONING DEBATE

The argument that cloning could allow infertile couples to have children doesn't hold up with opponents. "Infertility is not an absolute evil that justifies doing any and every thing to overcome it," Father Richard McCormick, a Jesuit and an ethicist at the University of Notre Dame, remarked to Kenneth L. Woodward of *Newsweek*. Others point out that couples who are not able to have their own biological children should consider the option of adopting a child who is already alive and in need of a home.

Some object to cloning for religious reasons. In a *Reader's Digest* article entitled "The Real Message of the Millennium," Paul Johnson compared the twentieth century's attempts at "social engineering, the practice of shoving large numbers of human beings around as though they were earth or concrete" to what he foresees as the twenty-first century's focus on "cloned humans, 'designer babies' and other alarming demonstrations that man now has the power to play God with lives." Asserting that social engineering was a "key feature in the Nazi and Communist totalitarian regimes" and was in the end defeated by Christianity, Johnson predicts that genetic engineering, including cloning, will also fail through the influence of Christianity: "Against this scientific background it is comforting to remember that Christianity, with its central message of submission to a higher being, remains so strong and vocal. The words of Jesus created a body of faith and morality that enabled humankind to defeat social engineering, and today it provides defenses against the threat of genetic engineering."

Some people do not believe that cloning itself is evil, but they have little faith that humans—subject as we are to greed and vengefulness and selfishness—are worthy of having such power. They fear that the cloning of animals, despite its potential to help the human race, would inevitably lead to the cloning of humans, a responsibility they believe we are not highly evolved enough to handle. "In science, the one rule is that what can be done will be done," Rabbi Moses Tendler, a professor of medical ethics, told Kenneth L. Woodward of *Newsweek*. As Sharon Schmickle pointed out in the *Minneapolis Star Tribune*, "Most scientists agree that much

---

WHAT IS EUGENICS?

Webster's defines eugenics as "a science that deals with the improvement of hereditary qualities of a race or breed." Eugenics is a hotly debated topic because it has often been used as a racist tool. Adolf Hitler believed that Germans were the ideal example of humankind, and he set out to eliminate "undesirable" races of people—predominantly Jews, but many others as well. The result was the Holocaust, which brought about the deaths of some six million people during World War II. This attempt at creating a "master race" is probably the best-known example of eugenics at work, and although it failed, many people today, known as white supremacists, still subscribe to it.

GENETIC ENGINEERING: THE CLONING DEBATE

more work is needed before the technology can be translated to humans. Still, it's noteworthy that the Roslin Institute in Edinburgh, Scotland [creators of Dolly the sheep], has moved to patent the cloning process, including its application to humans." Even the journal that broke the news of Dolly expects to see human clones someday. "When it published the paper reporting the birth of Dolly, the journal *Nature* wrote that it would not be surprised to see a human cloned within ten years," Sharon Begley reported in *Newsweek*.

Opponents such as ethicist Daniel Callahan believe that some standards of our society that are morally questionable but nonetheless already accepted might encourage people to accept cloning without fully considering the ramifications. "In our society there are two values which will allow anyone to do whatever she wants in human reproduction," Callahan told Woodward. "One is the nearly absolute right to reproduce—or not—as you see fit. The other is that just about anything goes in the pursuit of improved health."

Some people object to what they see as the commercial potential of cloning. Said *Newsweek*'s Sharon Begley: "Everyone in the cloning game sees dollar signs beckoning." Richard Seed is not the only one hoping to establish a cloning clinic. Begley noted: "These successes, preliminary as they are, have inspired a religious organization called the Raelian Movement to establish a company—Clonaid—to produce human clones for infertile or gay

THE ETHICS OF CLONING

couples, singles, or anyone who wants a genetic duplicate. Projected cost: about $200,000." Begley points out that the Raelians believe that humans were created by aliens from outer space.

Advocates for Further Research

There are some people who are not necessarily in favor of human cloning but who believe research in the area should continue, or at least that there should be no legal restrictions on such research. Many scientists oppose a cloning ban because they fear that such a ban would mean the end of new discoveries in genetic engineering. "Many scientists are urging policy makers to allow research to go forward regardless of restrictions that

The Raelian Movement believes that humans were created by aliens from outer space.

might be placed on its use. They envision cloning human cells and other applications that stop well short of copying a whole person," Sharon Schmickle reported in the *Minneapolis Star Tribune.*

Since cloning is so closely related to many other areas of genetic engineering and reproductive science, many researchers believe that a ban on cloning or cloning research could also cause a shutdown on work in these similar areas. Alta Charo of the National Bioethics Advisory Commission remarked to Ron Seely in the *Wisconsin State Journal* that Richard Seed's plan to open a cloning clinic could have an adverse effect on further research: "Unfortunately, the congressional proposals to date have been broad enough to ban valuable areas of scientific research that do not involve cloning to make babies. Ironically, Seed's announcement might tighten the noose around cloning, including valuable forms that do not involve human cloning."

The Concept of Eugenics

One question that everyone must face, whether they are for cloning, against cloning, or in favor of further research, is the morality of eugenics. Since cloning is, by definition, the practice of eugenics, a discussion of it does not fit into just one of the above three categories.

Eugenics becomes the central issue of the cloning debate when the focus moves from the cloning of animals

THE ETHICS OF CLONING

to the cloning of people. Whether they are for or against cloning of animals, most people do not see it as an issue of individual freedom for animals. The practice of a farmer's breeding certain animals to increase desirable traits in his herd is not seen as a violation of the animals' right to be unique (animals have no such inherent rights); it is simply the farmer's attempt to improve the quality of his livestock. Animals that are grown for food are not generally seen as having a right to life—if they're not good enough for breeding, they will be used as food, and if they have some defect that makes them unfit to eat, they will be destroyed. (The question of whether animals should be grown for food at all is another hotly debated topic; space does not permit a discussion of it here.)

A similar, though not identical, thought process can be applied to the breeding of domesticated animals. Purebred dogs and cats are prized by many people, who enter them in shows, or value them for traits specific to their breed, or simply consider them status symbols. The difference in the way most people think about such animals becomes obvious when a "defective" one is born. Unlike agricultural animals, those that do not turn out "perfect" are not simply eaten instead of bred. So what happens to them? While most people do not care about the killing of agricultural animals, they would be outraged to hear that a domesticated animal had been killed

GENETIC ENGINEERING: THE CLONING DEBATE

because it had the wrong eye color or didn't hold its tail correctly. Although most people would agree that a domestic animal has a right to live even if it is not show quality, no one would argue that a dog or cat with less than perfect traits has the inherent right to breed. In fact, the vast majority of people do not recognize that any domestic animal has an inherent right to reproduce, whether it is a purebred or not. The exact opposite—the neutering of pets—is considered to be the humane, intelligent choice for pet owners.

Selective breeding for desirable traits, as in purebred dogs, is an aspect of eugenics.

So, agricultural animals have neither the right to breed nor the right to live if their breeding and/or living does

THE ETHICS OF CLONING

not suit the farmer's purposes, and domestic animals are believed to have a basic right to live but still do not have the inherent right to breed. Now what about people?

In the United States, the Constitution provides for every human being's basic right to life. It is also generally accepted that all people have the right to have children; whether or not they are fit parents is a separate issue. These rights apply to everyone (unless they have been convicted of a crime and sentenced to death or have been involuntarily sterilized), and it is these rights that people fear the practice of eugenics will violate.

When a couple decides to have a baby, if they hope for a boy instead of a girl, or if they hope that the baby will have its mother's beautiful blue eyes and not its father's big nose, that is eugenically inspired thinking. If they find, partway through the pregnancy, that the baby has a genetic defect and they decide to abort it, that is a eugenic decision—a decision related to the production of good offspring. People who decide not to have children at all because they know they carry a serious genetic defect and are likely to pass it on to their offspring are making a eugenic decision. But the science of eugenics, which deals with the improvement of the hereditary qualities of an entire race or breed, is something else altogether. Protests often arise when people use eugenics in an attempt to change others, or to alter an entire population, or to engineer a perfect race.

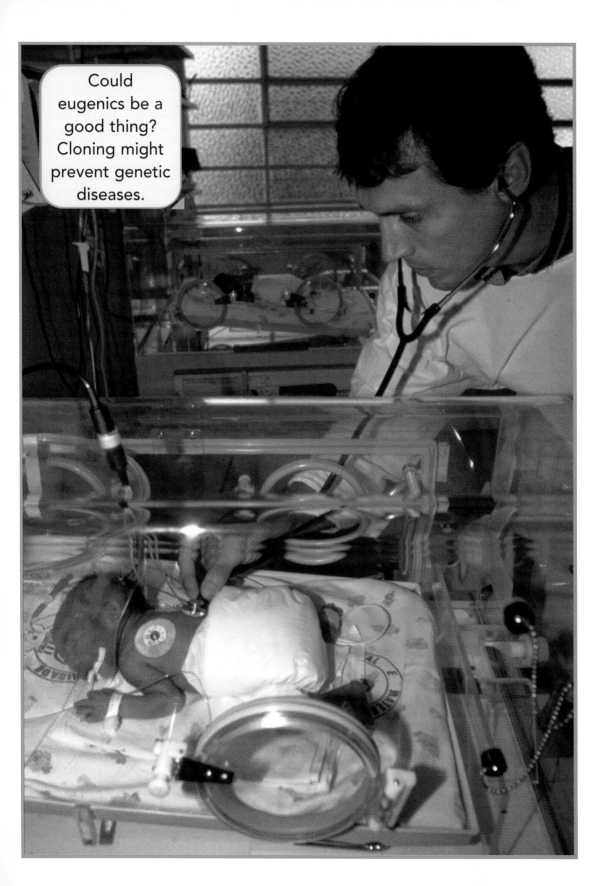

THE ETHICS OF CLONING

This is where cloning comes into the debate. The simple question is, "Who will decide who gets cloned?" Again, if it is a couple who wants a child and they clone one of themselves, that would most likely be seen as a personal decision and not an assault on everyone else's freedom. But what if poor people were discouraged from cloning themselves? This would likely happen if human cloning became commonplace because only people who could afford it would be able to do it. And what if people were not allowed to use cloning as another artificial form of reproduction, but scientists or the government decided that certain people should be cloned for the good of society as a whole? Who would those people be? And what if they didn't want to do it? Suppose cloning became common, and scientists began using it to eradicate genetic diseases. Would people who carried defective genes still be allowed to have children? These are the kinds of questions asked by people who fear eugenics and who fear the use of cloning as its tool.

Others believe that, exercised responsibly, eugenics is the answer to much of the suffering in the world today. If it could eliminate some diseases and prevent babies from being born into a life of pain, or if it could help people to live longer, stronger lives, some believe eugenics would be a good thing. Some of these supporters of eugenics also believe that it could be controlled and not used for evil purposes if only we as a society would work toward that goal.

SHOULD CLONING CONTINUE?

As you can see from what you have read so far, there are hundreds of questions surrounding the issue of cloning. What is your opinion? Do you think cloning should continue? Consider what you have read in the preceding chapters, along with other reading or research you've done and discussions you've had with friends, your parents, or in school. Then use the following questions to further explore the facts, opinions, and issues involved in this debate. After you've learned even more about the topic of cloning, you may want to use your knowledge to write a research paper or create a presentation for your science, history, ethics, government, or religion class.

Bacteria versus Animals versus Humans

If someone believes cloning is right or wrong, should that person have to hold that opinion for all forms of cloning?

SHOULD CLONING CONTINUE?

Would it be inconsistent for a person to accept cloning of simpler life forms such as bacteria but to reject it for humans? Or is it OK for a person to accept some forms of cloning and not others? If you believe cloning is acceptable for some organisms but not others, how would you propose keeping cloning limited only to the organisms for which you believe it is OK?

Does It Depend on the Reason?

Should the question of whether or not cloning should continue be based on the goals behind it? Do you believe that some reasons for cloning are more worthy or valuable than others? If this became the way it was decided whether certain people could make clones, who would monitor the work? Who would make sure that it was being done only for noble reasons, not for illegal or immoral ones?

What to Do with the Failed Attempts

Imagine that there were 277 attempts to clone a human and only one worked, as was the case with Dolly the sheep. What should happen to the 276 others? Would they be just clumps of cells that didn't "take," or babies that are born with birth defects? Is it right to make the attempt to clone a person if the chances of it working are only .0036 percent? Is it right to risk creating children with birth defects

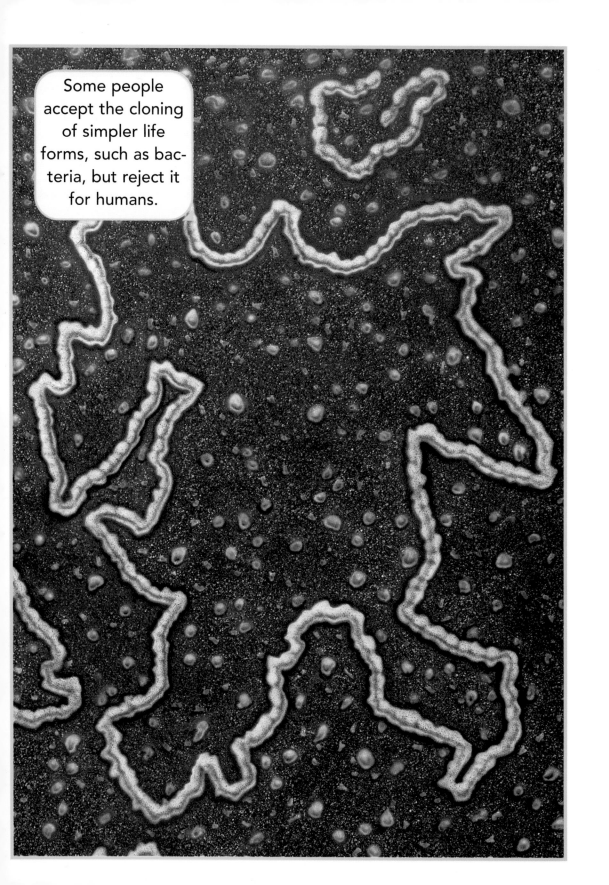

while trying for a "good" one? What would you do with the failures—abort them, or destroy them after they are born, or make arrangements for them to be cared for after birth?

Some people believe that our society's continued advancements in biotechnology will lead to discrimination against people who do not abort babies with birth defects. How do you feel about that prediction? How would society treat the babies who are born with birth defects due to imperfect cloning attempts?

What If Cloning Could Save Your Life?

Some people advocate the development of human cloning for its potential to provide plenty of organ donors. What would be done with the rest of the person once the organs were harvested?

ETHICAL CONSIDERATIONS

If you had a fatal disease that could be cured only by a bone marrow transplant, but no one was a match for you, would you want to be cloned in order to create another person who could serve as a bone marrow donor for you? Suppose you are only eight years old; would your clone be your sibling (your parents' child) or your own child? What if you found out that the baby was going to be born with painful deformities and mental retardation, but also with the bone marrow you needed; would you still want it to be born?

GENETIC ENGINEERING: THE CLONING DEBATE

Should people be "grown" with the intention of using their organs to cure others? Would it be OK only for organs such as kidneys, where one could be taken without killing the donor? What about organs like hearts, where the donor would obviously have to die for the organ to be given to someone else? If such donor people were created, where would they live? Would they have parents? What impact would they have on society's beliefs about discrimination?

Cloning As Artificial Reproduction

Do you think that people who are unable to have children should be given the opportunity to use cloning to create a child? What if they can't afford to pay for it? What about the option of adopting a child? Does your answer depend on whether the people are a married couple or a homosexual couple? Does it depend on whether the person to be cloned has any genetic defects that could possibly cause the child to eventually develop a disease? Since the child would be, genetically, the product of only one person, would that person's spouse or partner have any claim to the child as his or her own?

Suppose a child was created by cloning. What advantages or disadvantages do you think he or she might have? How would that child be treated in school? By

SHOULD CLONING CONTINUE?

Should cloning become an alternative to adoption for people who can't have children?

society? By his or her parents? How would you react to a new brother or sister who was cloned from one or the other of your parents?

Cloning As Duplication of Another Person

How do you feel about the opinion that a grieving parent whose child has just been killed might be comforted by cloning that child? Would the parent be trying to re-create the same child? If so, considering what you have read in this

GENETIC ENGINEERING: THE CLONING DEBATE

book, would that parent succeed or fail? What are the psychological implications for the parent in this situation? How do you think the child would feel about being born as a "replacement" for someone else? Would he or she feel obligated to be the other person?

Do you believe a clone would be a real, true person?

Should It Be Free?

If human cloning does become a reality, should it be available to everyone, regardless of whether they have the money to pay for it? If so, why, and if not, why not? Would it depend on what the clone would be used for? Who would decide whether a person's reason for wanting a clone was good enough?

Should cloning research be funded by the government? If so, how much control should the government have over what aspects of cloning are studied, who has access to the information, and how it is used? If not, who should pay for it, and how would it be monitored to prevent abuses?

Should It Be Available to Anyone?

Aside from the question of affordability, should cloning be allowed for anyone who wants it? What if someone like serial killer Ted Bundy or mass murderer Charles

Manson decided he wanted to be cloned? What if the entire Ku Klux Klan, a group that advocates white supremacy, decided to make clones of each of its members in order to form a bigger, stronger group with which to discriminate against nonwhites? What if a group of people decided to overthrow the government by cloning its own army? Or a small foreign country decided to clone its residents in order to strengthen itself and become a superpower?

Who would decide whether or not people would be allowed to clone themselves? Would there be an application process? Would you need a permit, similar to a gun permit? Consider the fact that there are no such requirements for people who have children the natural way. Would countries be allowed only a limited number of clones?

> **WHAT ABOUT RELIGION?**
>
> Whether or not you practice a religion, do you believe that religions should have a say in whether or not cloning continues? Do you think theology, or religious thought, has any relevance in debates over the ethics of genetic engineering?
>
>

GENETIC ENGINEERING: THE CLONING DEBATE

Cloning and Religion

If you practice a religion, what is your religion's stance on cloning? What questions does it ask? Has it made a declaration either for or against cloning or genetic engineering in general? If your religion recognizes God or another supreme being, what do you think he or she would say about cloning?

If your religion believes in reincarnation, what do you think a cloned person would have been in a past life? What about the theory of karma, which states that a person's past life determines what he or she will be reborn as—what behavior would lead to a person being reborn as a clone?

Cloning raises many questions for religious people.

SHOULD CLONING CONTINUE?

Would only the most important people be reborn this way, or those who had led the most honorable lives, or would it be a punishment?

Would clones have souls? Would they go to heaven or another form of afterlife when they died?

How Does Eugenics Fit In?

How do you feel about eugenics? Is it right or wrong, or does your answer depend on the circumstances in which it is used? Do you think it would be good to clone the best people in order to create more people like them? Who would decide whom to clone?

What Hasn't Been Thought of Yet?

What other uses can you think of for cloning? What other ways could it be beneficial to humankind or to individuals? What other abuses could it be used for? Do you think enough thought has been given to these issues to allow cloning research to continue safely?

Conclusion

Cloning is a topic that is highly controversial and one that will not just go away. Many of the debates, arguments, and facts related to this issue have been presented here, but even more could be written. While much of the decision

GENETIC ENGINEERING: THE CLONING DEBATE

making will be done by legislators and scientists, individuals like you do have a say in whether or not cloning should continue. If you want your opinion on cloning to be heard, write an editorial for your school and local newspapers; contact your state and national legislators and let them know what you think; find out which schools and research centers are doing work in cloning and write to them as well.

GLOSSARY

artificial reproduction   Any method of conceiving a child that involves outside help, such as drugs, the use of a surrogate mother, or genetic engineering.

bioethics   The study of the ethical implications of scientific experimentation.

chimera   A creature made up of genetic material from more than one species (either more than one type of animal or a human-animal combination).

chromosome   The part of a cell that contains the genes.

cloning   Creating a twin through genetic engineering.

DNA   Deoxyribonucleic acid, one of the nucleic acids responsible for inherited characteristics in an organism.

ecosystem   The combination of animals and their environment, each of which is affected by the other.

embryo   A developing organism that has not yet been born.

eugenics   A science that deals with the improvement of the hereditary qualities of a race or breed.

GENETIC ENGINEERING: THE CLONING DEBATE

gene   The part of a cell that controls inherited characteristics such as eye color.

genetic engineering   Changing an organism's genetic makeup in order to strengthen desired qualities or remove undesired ones. Also, taking genetic material from one organism to create a new one.

infertile   Unable to have children through natural methods.

monoculture   Usually refers to farm crops that are all the same type of plant, or a crop in which all the plants have come from the same genetic beginnings.

somatic cell   A cell that is on its way to becoming a certain type of organ or tissue.

somatic cell nuclear transfer   A method of cloning in which the nucleus of one cell is removed and used to replace the nucleus of another cell.

surrogate mother   One who carries and delivers a child for someone else who is not able to; usually, the child is not genetically her own.

telomere   The segment of a chromosome believed to be related to aging in an animal.

totipotent cell   A cell that has not yet become somatic and is therefore ideal for use in cloning because it has the potential to become any part of the body.

xenotransplantation   The transplanting of organs from one type of animal to another or from animals to humans.

FOR FURTHER READING

Andrews, Lori B. *The Clone Age: Adventures in the New World of Reproductive Technology*. New York: Holt, 1999.

Cole-Turner, Ronald, ed. *Human Cloning: Religious Responses*. Louisville, KY: Westminster/John Knox Press, 1997.

Gosden, Roger. *Designing Babies: The Brave New World of Reproductive Technology*. New York: W. H. Freeman & Co., 1999.

Huxley, Aldous. *Brave New World*. New York: Harper & Row, 1932.

Kas, Leon R., and James Q. Wilson. *The Ethics of Human Cloning*. Washington, DC: AEI Press, 1998.

Kolata, Gina Bari. *Clone: The Road to Dolly, and the Path Ahead*. New York: Morrow, 1998.

McGee, Glenn, editor. *The Human Cloning Debate*. Berkeley, CA: Berkeley Hills Books, 1998.

GENETIC ENGINEERING: THE CLONING DEBATE

Nussbaum, Martha, and Cass R. Sunstein, eds. *Clones and Clones: Facts and Fantasies about Human Cloning.* New York: W. W. Norton & Co., 1998.

Pence, Gregory E., ed. *Flesh of My Flesh: The Ethics of Cloning Humans*. Larham, MD: Rowman & Littlefield, 1998.

Silver, Lee M. *Remaking Eden: How Genetic Engineering and Cloning Will Transform the American Family.* New York: Avon, 1998.

FOR MORE INFORMATION

ORGANIZATIONS AND WEB SITES

Council for Responsible Genetics
5 Upland Road, Suite 3
Cambridge, MA 02140
(617) 868-0870 phone
(617) 491-5344 fax
Web site: http://www.gene-watch.org
E-mail: crg@gene-watch.org

Publishes a newsletter entitled *GeneWatch* (not related to the organization GeneWatch UK). Describes itself as "a national nonprofit organization of scientists, public health advocates, and others which monitors the development of new genetic technologies and advocates for their responsible use."

Eureka!Science
http://www.eurekascience.com

GENETIC ENGINEERING: THE CLONING DEBATE

Web site geared toward young people that explains how genetic engineering and cloning work.

GeneWatch UK
The Courtyard
Whitecross Road
Tideswell
Buxton
Derbyshire SK17 8NY
United Kingdom
+ 44 (0) 1298 871898 phone
+ 44 (0) 1298 872531 fax
Web site: http://www.genewatch.org
E-mail: gene.watch@dial.pipex.com

A British group that describes itself as "an independent organization concerned with the ethics and risks of genetic engineering. It questions how, why, and whether the use of genetic technologies should proceed and believes that the debate over genetic engineering is long overdue." Web site contains information on genetic engineering and a list of links to other sites.

Kennedy Institute of Ethics
Georgetown University
Box 57212
Washington, DC 20057-1212
(202) 687-8099 phone

FOR MORE INFORMATION

(202) 687-8089 fax
Web site: http://www.georgetown.edu/research/kie
E-mail: kicourse@gunet.georgetown.edu

National Bioethics Advisory Commission
6100 Executive Boulevard, Suite 5B01
Rockville, MD 20892-7508
Web site: http://bioethics.gov
    Web site of the group of people appointed by President Clinton to determine what should be done about cloning research.

Students for Alternatives to Genetic Engineering (SAGE)
Web site: http://www.sage-intl.org
E-mail: sage@expertsforum.com
    A group of students who propose options other than cloning and genetic engineering.

University of Virginia Center for Biomedical Ethics
Web site: http://www.med.virginia.edu/bioethics
    Another source for discussion of the ethical implications of cloning.

INDEX

A
abortion, 41, 47
agricultural animals, 32, 39–41
American Society for Reproductive Medicine, 19
animal hybrids, 12

B
Begley, Sharon, 36–37

C
Callahan, Daniel, 36
cell line, 10–11, 26
cells
    differentiated, 9
    somatic, 9–10
    totipotent, 9
Charo, Alta, 21–22, 38
chimeras, 12
Christianity, 34–35

Clinton, President, 16–17, 22
cloning
    animal, 13–19
    arguments against, 29–37
    arguments for, 24–29
    arguments for further research, 37–38
    as artificial reproduction, 27, 31, 43, 48
    eugenics and, 38–43, 53
    funding for, 17, 23, 50
    human, 8, 11, 13, 15, 16, 17, 20–23, 26, 32, 36, 37–38, 43
    questions concerning, 44–53
    religion and, 34–35, 52–53
Council for Responsible Genetics, 31–32
Cumulina, 14
cystic fibrosis, 26

INDEX

D
diabetes, 26
DNA, 7, 13, 16
Dolly, 13, 14, 15–16, 17, 36, 45
domesticated animals, 39–41
drug testing, 25

E
eugenics, 38–43, 53
extinction of species, prevention of, 29

F
farming, 12, 39, 41
Fibro, 14, 15

G
genetic engineering, 7, 11–12, 22, 25, 34–35, 37, 38, 52

H
hemophilia, 26

I
identical twins, 8
infertility, 20, 27, 32–34, 36

J
Johnson, Paul, 34

M
McCormick, Father Richard, 34
monoculture, 32

N
National Bioethics Advisory Commission, 17–19, 22, 26, 38
natural selection, 11–12

O
organ donation/transplants, 25–26, 47–48

P
Prolume Ltd., 26–27

R
Raelian Movement, 36–37
Roslin Institute, 15–16, 35–36

S
Schmickle, Sharon, 31, 35–36, 38
Seed, Richard, 20–22, 36, 38
Seely, Ron, 17, 22, 38
"social engineering," 34
somatic cell nuclear transfer, 9–10, 15

T
telomeres, 16
Tendler, Rabbi Moses, 35

W
Wilmut, Ian, 15–16
Woodward, Kenneth L., 34, 35, 36

X
xenotransplantation, 25, 26

GENETIC ENGINEERING: THE CLONING DEBATE

About the Author
Debbie Stanley has a B.A. in journalism and an M.A. in industrial and organizational psychology. She lives in New Baltimore, Michigan.

Photo Credits
Cover © C. Raymond/Science Source/Photo Researchers Inc.; p. 2 © Geoff Tompkinson/ Science Photo Library/Photo Researchers Inc.; p. 6 © Digital Art/CORBIS; p. 10 © James Holmes/Celltech Ltd./Science Photo Library/Photo Researchers Inc.; p. 12 © Lowell Georgia/Science Source/Photo Researchers Inc.; p. 14 © Reuters/HO/Archive Photos; p. 17 © Reuters/Peter Morgan/Archive Photos; p. 18 © David Parker/Science Photo Library/Photo Researchers Inc.; p .21 © Reuters/Sue Orgrocki/Archive Photos; p. 25 © Anronia Reeve/Science Photo Library/Photo Researchers Inc.; p. 28 © Popperfoto/Archive Photos; p. 30 © James King-Holmes/Science Photo Library/Photo Researchers Inc.; p. 33 © James P. Blair/CORBIS; p. 37 © Warner Bros./Everett Collection; p. 40 © Kevin R. Morris/CORBIS; p. 42 © Jon Spaull/CORBIS; p. 46 © Francis Leroy, Biocosmos/Science Photo Library/Photo Researchers Inc.; p. 47 © Simon Fraser/Department of Haematology, RVI, Newcastle/Science Photo Library/Photo Researchers Inc.; p. 49 © Dusty Willison/ International Stock; p. 51 © James L. Amos/CORBIS; p. 52 © Jules T. Allen/CORBIS.

Series Design
Mike Caroleo